Science in the News
Critical Thinking Worksheets

Physical Science

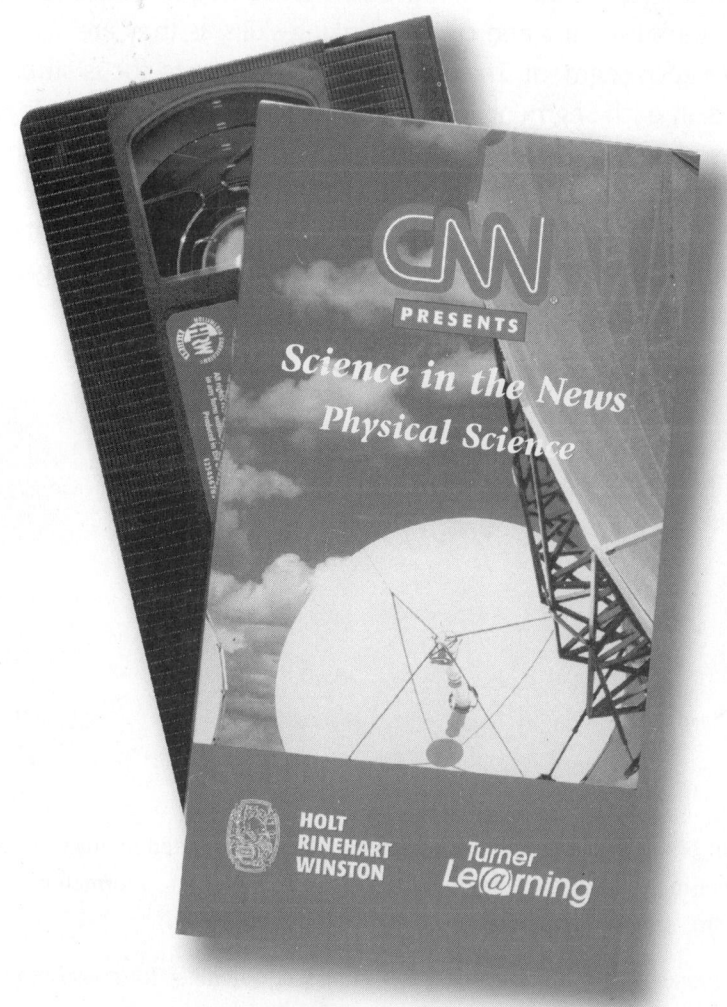

HOLT, RINEHART AND WINSTON
A Harcourt Classroom Education Company

Austin • New York • Orlando • Atlanta • San Francisco • Boston • Dallas • Toronto • London

To the Teacher

The video segments in the *CNN Presents Science in the News: Physical Science* program bring the knowledge and experience of the CNN news team right into your classroom. Show students the relevance of physical science to everyday life with interesting news segments that relate to the topics they are studying in class. The segments may be used as warm-up activities to start the class, as diversion points to stimulate class discussion, or as springboards for further research. Your students will see the efforts of chemists, environmentalists, students, governments, and many others who are using science to answer important questions.

These Critical Thinking Worksheets contain thought-provoking questions that require students to carefully examine the information presented in each segment. Students can sharpen their listening and critical thinking skills as they are challenged to evaluate each segment. The worksheets also help to focus students' attention on the topics in each segment. You may wish to have your students use their textbook as a resource for answering questions. Also, topics covered in the segment can be used as the basis for student-led discussions. The Teacher's Guide that accompanies the *Physical Science* video contains discussion points, research ideas, and other information that will help you and your students get the most out of these worksheets.

Copyright © by Holt, Rinehart and Winston

All rights reserved. No part of this publication may be reproduced or transmitted in any form or by any means, electronic or mechanical, including photocopy, recording, or any information storage and retrieval system, without permission in writing from the publisher.

Teachers may photocopy worksheets in complete pages in sufficient quantities for classroom use only and not for resale.

Printed in the United States of America

ISBN 0-03-056554-5

3 4 5 6 7 8 9   862   04 03 02 01

Contents

	Segment Title	Page Number
Segment 1	Amusement Park Physics	1
Segment 2	Land Speed Record	2
Segment 3	Trebuchet Design	3
Segment 4	Crash-Test Dummies	4
Segment 5	Smart Skin Sensors	5
Segment 6	Egg Drop Contest	6
Segment 7	Zero-Gravity Plane	6
Segment 8	Circus Acrobats	7
Segment 9	Turbulent Flow	7
Segment 10	Wet Design	8
Segment 11	Energy-Saving House	9
Segment 12	Urban Heat Islands	10
Segment 13	Water-Cooled City	10
Segment 14	Virtual Practice Room	11
Segment 15	Japanese Telescope	12
Segment 16	Color-Deficiency Lenses	13
Segment 17	Holograms	14
Segment 18	Fusion Lighting	15
Segment 19	Edison's Lab	16
Segment 20	Eagle Electrocution	17
Segment 21	Magnetic Attractions	18
Segment 22	Atom Laser	19
Segment 23	Portable Solar Power	20
Segment 24	Wisp of Creation	21
Answer Key		**23**

Name _____ Date _____ Class _____

Science in the News: Critical Thinking Worksheets

Segment 1

Amusement Park Physics

1. In which activities in the segment do students experience centrifugal force?

   _____
   _____
   _____
   _____
   _____
   _____

2. A student in the video says, "We learned that this ride pulls a lot of Gs on your body." What does the phrase "pulling Gs" mean?

   _____
   _____
   _____
   _____
   _____
   _____

3. Describe how forces must act on an object to move the object in a circular path.

   _____
   _____
   _____
   _____
   _____

Name _____ Date _____ Class _____

Science in the News: Critical Thinking Worksheets

Segment 2
Land Speed Record

1. The reporter says that ceramic coatings are needed because "at speeds above 3400 mi/h, even the hardest steel can melt." How can high speeds melt steel? Write a more precise statement explaining why ceramic coatings are needed.

   _____

   _____

   _____

   _____

   _____

   _____

   _____

2. Why is a helium-filled tent used in the last 2 mi of the run?

   _____

   _____

   _____

   _____

   _____

   _____

3. If the sled and the attached rocket have a combined mass of 2300 kg, what force is required to achieve an acceleration of 516 m/s^2? Show your calculation in the space below.

Name _____ Date _____ Class _____

Science in the News: Critical Thinking Worksheets

Segment 3
Trebuchet Design

1. During the process of trying to draw plans for a trebuchet based on old illustrations, what problem had to be solved?

   _____
   _____
   _____
   _____
   _____

2. The basket of a trebuchet is attached 2.7 m from the pivot, and the object to be hurled is attached 7.4 m from the pivot. Assume that there is no appreciable friction. Calculate the mechanical advantage of the trebuchet in the space below.

3. At a trebuchet competition, one contest is to see who can come closest to a target placed 250 m away on the ground. Your trebuchet launches a 35.0 kg stone, and the stone reaches a maximum height of 140.0 m. What is the stone's gravitational potential energy at that height? Calculate the stone's kinetic energy when it hits the ground at a downward velocity of 52.4 m/s. Show your calculations in the space below.

4. How are the values for potential energy and kinetic energy that you calculated in item 3 related?

   _____
   _____
   _____
   _____

Science in the News: Critical Thinking Worksheets   Physical Science **3**

Name _____ Date _____ Class _____

Science in the News: Critical Thinking Worksheets

Segment 4
Crash-Test Dummies

1. What requirements did air bags have to meet at the time of this video? What changes are being proposed?

   _____
   _____
   _____
   _____
   _____
   _____

2. All air bags have large holes in them. You can see the holes in one of the slow-motion sequences in the video. Why do you think these holes are needed, and why do they not reduce the effectiveness of the air bag? (Keep in mind that to protect a car's occupants, the air bags must be able to exert a large force for a short period of time.)

   _____
   _____
   _____
   _____
   _____
   _____

3. How does the motion of a car's occupants during a crash illustrate Newton's first law?

   _____
   _____
   _____
   _____
   _____
   _____

Name_____ Date_____ Class_____

Science in the News: Critical Thinking Worksheets

Segment 5

Smart Skin Sensors

1. Under what conditions are people most likely to be injured by automobile air bags?

   _____
   _____
   _____
   _____

2. What information does the smart skin sensor send to the car's computer?

   _____
   _____
   _____
   _____

3. Describe the three different ways that an air bag could respond to information from the sensor.

   _____
   _____
   _____
   _____

4. Explain how the sensor could help prevent airbag–related deaths and injuries, such as those discussed in Segment 4.

   _____
   _____
   _____
   _____
   _____

Science in the News: Critical Thinking Worksheets   Physical Science

Name _____ Date _____ Class _____

Science in the News: Critical Thinking Worksheets

Segment 6

Egg Drop Contest

1. The video shows two different strategies that students use to get the egg to survive the fall. Identify and describe these strategies.

   _____
   _____
   _____

2. Suppose you drop an egg that has a mass of 68 g from a height of 15 m, and the egg reaches a downward velocity of 17 m/s. What momentum would the egg have at the moment it hits the ground? Show your calculation in the space below.

Segment 7

Zero-Gravity Plane

1. Although the reporter corrects himself later, he incorrectly states that the purpose of the flight is for students to "get a taste of life without gravity." What exactly are the students experiencing?

   _____
   _____

2. Describe the motion that the plane must have to produce a "weightless" condition.

   _____
   _____

3. Describe what the students experience when the plane levels off at the bottom of the dive. Explain the sensation in terms of inertia and acceleration.

   _____
   _____
   _____

Name_____ Date_____ Class_____

Science in the News: Critical Thinking Worksheets

Segment 8
Circus Acrobats

1. The rotational inertia of a spinning object depends on the mass and diameter of the object. Which of these factors do the acrobats change to manipulate their motion?

   _____
   _____

2. Is it easier to do a somersault in a tucked or a fully extended position? Explain why.

   _____
   _____
   _____
   _____

Segment 9
Turbulent Flow

1. Describe the motion of the 737 jet shown early in the segment. What is causing this motion?

   _____
   _____
   _____
   _____

2. List two advantages of using soap films to study turbulence.

   _____
   _____
   _____
   _____

Name_____ Date_____ Class_____

Science in the News: Critical Thinking Worksheets

Segment 10
Wet Design

1. How does the water gun create sudden, high spurts of water?

   _____
   _____
   _____
   _____
   _____

2. What three criteria must be met by the fountains the company designs?

   _____
   _____
   _____
   _____
   _____

3. The water in the segment is described as having axisymmetric laminar fluid flow. Find out the meaning of *laminar flow* and how this type of flow differs from turbulent flow. Summarize your findings in the space below.

   _____
   _____
   _____
   _____
   _____
   _____
   _____

Name _____ Date _____ Class _____

Science in the News: Critical Thinking Worksheets

Segment 11

Energy-Saving House

1. The range burners shown probably consume electrical energy at a rate of 2.0 kW or more. How can you account for the suggestion that 1.7 kW can power all of the appliances in the house?

   _____
   _____
   _____
   _____

2. Contrast the operation of the geothermal heat pump in summer with its operation in winter.

   _____
   _____
   _____
   _____

3. Most windows in the house face south. How does this arrangement help heat the house in winter? What structure helps keep the house from overheating in summer? What factor explains why this arrangement works both in summer and in winter?

   _____
   _____
   _____
   _____

4. Why do power companies offer incentives to construct energy-saving homes?

   _____
   _____
   _____
   _____

Name_____ Date_____ Class_____

Science in the News: Critical Thinking Worksheets

Segment 12
Urban Heat Islands

1. What factors cause cities to become heat islands? The reporter attributes this problem to increasing population, but people do not radiate enough heat to cause the problem. How could the cause have been stated more precisely?

   _____
   _____
   _____
   _____

2. What are people doing to overcome the heat-island effect?

   _____
   _____
   _____
   _____

Segment 13
Water-Cooled City

1. What property of water causes Lake Ontario to remain cold in summer while the surrounding land becomes hot?

   _____
   _____
   _____

2. What does the city of Toronto hope to gain by using the lake water?

   _____
   _____
   _____

Name _____ Date _____ Class _____

Science in the News: Critical Thinking Worksheets

Segment 14
Virtual Practice Room

1. Take note of the sound of the trumpet player in the practice room. Describe how the sound of her playing when the system is operating differs from the original sound.

   _____
   _____
   _____
   _____
   _____
   _____

2. What are reverberations? How do the reverberations in a large arena differ from those in a smaller arena?

   _____
   _____
   _____
   _____
   _____
   _____

3. For what reasons might a school install such practice rooms?

   _____
   _____
   _____
   _____
   _____
   _____

Name _____ Date _____ Class _____

Science in the News: Critical Thinking Worksheets

Segment 15

Japanese Telescope

1. Of the three types of mirrors—concave, flat, and convex—which type must be used as the telescope's main mirror so that it has a main focal point? Explain why the other two types of mirrors would not work.

   _____
   _____
   _____
   _____
   _____
   _____
   _____
   _____
   _____

2. Why could looking at distant stars and galaxies yield clues about the origin of life?

   _____
   _____
   _____
   _____
   _____
   _____
   _____
   _____

Name_____ Date_____ Class_____

Science in the News: Critical Thinking Worksheets

Segment 16

Color-Deficiency Lenses

1. What are the two kinds of light-sensitive cells in the retina of the eye? Describe the function of each type.

   _____
   _____
   _____
   _____
   _____
   _____

2. The reporter states incorrectly that cones "pick up and then transmit one of three colors—red, blue, or green—sending each to the brain in the form of wavelengths." Replace this part of the script by writing a more accurate description of the mechanism of color vision.

   _____
   _____
   _____
   _____
   _____
   _____

3. Why is it possible to see a full-color image when the cones in the retina are only sensitive to the colors red, blue, and green?

   _____
   _____
   _____
   _____
   _____

Name_____ Date_____ Class_____

Science in the News: Critical Thinking Worksheets

Segment 17
Holograms

1. What colors are the lasers used in the laboratory? Why are three lasers necessary?

_____
_____
_____
_____
_____
_____
_____

2. Why are the lasers mounted on a floating table?

_____
_____
_____
_____
_____
_____

3. Why is the Cartier company interested in these full-color holograms?

_____
_____
_____
_____
_____
_____
_____

Name_____ Date_____ Class_____

Science in the News: Critical Thinking Worksheets

Segment 18

Fusion Lighting

1. In what ways does the bulb shown in the video differ from conventional light bulbs and fluorescent lights?

   _____
   _____
   _____
   _____
   _____

2. What is the basis for the claim that sulfur lamps will last much longer than ordinary lamps?

   _____
   _____
   _____
   _____

3. What economic benefit of fusion lighting is suggested in the video?

   _____
   _____
   _____
   _____

4. What is a plasma? How does it differ from a gas?

   _____
   _____
   _____
   _____

Name _____ Date _____ Class _____

Science in the News: Critical Thinking Worksheets

Segment 19
Edison's Lab

1. In what ways were people's lives affected by Edison's work?

   _____
   _____
   _____
   _____
   _____

2. List three of Edison's inventions.

   _____
   _____
   _____
   _____

3. The boy in the segment asserts that without Edison, we would not have electricity or television today. Do you agree or disagree with this statement? Explain your choice.

   _____
   _____
   _____
   _____

4. What is the importance of Edison's lab in the development of science and technology in the United States?

   _____
   _____
   _____
   _____

Name _____ Date _____ Class _____

Science in the News: Critical Thinking Worksheets

Segment 20
Eagle Electrocution

1. Why are electric lines that pass over poles and enter transformers hazardous for large birds but not for smaller birds? Explain your answer in terms of electric circuits.

   _____
   _____
   _____
   _____
   _____
   _____
   _____

2. What remedies are proposed to help solve this problem?

   _____
   _____
   _____
   _____
   _____
   _____
   _____

3. Why are transformers necessary? What purpose do they serve?

   _____
   _____
   _____
   _____
   _____
   _____
   _____

Name _____ Date _____ Class _____

Science in the News: Critical Thinking Worksheets

Segment 21
Magnetic Attractions

1. How might magnets be used to prevent pollutants from entering the environment?

   _____
   _____
   _____
   _____
   _____
   _____

2. According to the video, what is the difference between paramagnetism and diamagnetism? Give examples of objects that exhibit each type of magnetic behavior.

   _____
   _____
   _____
   _____
   _____
   _____

3. Explain in terms of forces how scientists are able to levitate objects.

   _____
   _____
   _____
   _____
   _____
   _____

Name _____ Date _____ Class _____

Science in the News: Critical Thinking Worksheets

Segment 22
Atom Laser

1. How does the laserlike atom beam differ from an actual laser?

   _____
   _____
   _____
   _____
   _____
   _____
   _____

2. Why must the atoms be cooled to near absolute zero?

   _____
   _____
   _____
   _____
   _____
   _____
   _____

3. What applications are likely for the atom laser in the future?

   _____
   _____
   _____
   _____
   _____
   _____
   _____

Name _____ Date _____ Class _____

Science in the News: Critical Thinking Worksheets

Segment 23
Portable Solar Power

1. Why is the solar photovoltaic generating system particularly suited to this research facility? Why do you think this system would probably not be as useful in northern California or in Oregon?

   _____
   _____
   _____
   _____
   _____
   _____

2. Why do the people in the video believe the photovoltaic panels are a better idea than connecting the stations to power lines that are farther than 5 mi away?

   _____
   _____
   _____
   _____
   _____
   _____

3. If the panels can produce 10.0 kW of electric power, what is the maximum current that can be delivered at a voltage of exactly 220 V? Show your calculation in the space below.

20 Science in the News: Critical Thinking Worksheets   Physical Science

Name _____ Date _____ Class _____

Science in the News: Critical Thinking Worksheets

Segment 24
Wisp of Creation

1. What elements does the video suggest were present after the big bang?

   _____
   _____
   _____
   _____
   _____

2. Why is it more significant to find evidence of helium in an object 10 billion light-years away than in a star much closer to Earth?

   _____
   _____
   _____
   _____
   _____
   _____
   _____

3. How is "dark matter" characterized by Dr. Davidsen?

   _____
   _____
   _____
   _____
   _____
   _____
   _____

Science in the News: Critical Thinking Worksheets Answer Key

Segment 1
Amusement Park Physics
1. Students experience centrifugal force as the roller coaster turns around curves.
2. A ride that "pulls Gs" exerts a force on the rider's body greater than the force of gravity. If a ride pulled 3 Gs, for instance, the rider would experience a force three times the force of gravity.
3. Forces must be directed toward a central point to move an object in a circular path.

Segment 2
Land Speed Record
1. High speeds can melt steel because of friction. An example of a better statement is "At speeds above 3400 mi/h, enough heat is produced by friction to melt even the hardest steel."
2. Helium is light and helps reduce the heat and friction caused by high velocity travel through Earth's atmosphere, boosting the sled's final speed.
3. 1.2×10^6 N
   $F = (2300 \text{ kg})(516 \text{ m/s}^2) = 1.2 \times 10^6$ N

Segment 3
Trebuchet Design
1. Old illustrations lacked details about important parts of the trebuchet, such as its bearings and axles.
2. 0.36
   $MA = (2.7 \text{ m})/(7.4 \text{ m}) = 0.36$
3. grav. PE = KE = 4.8×10^4 J
   grav. PE = $(35.0 \text{ kg})(140.0 \text{ m})(9.8 \text{ m/s}^2)$
   $= 4.8 \times 10^4$ J
   KE = $(1/2)(35.0 \text{ kg})(52.4 \text{ m/s})^2$
   $= 4.8 \times 10^4$ J
4. The values for potential energy and kinetic energy are equal.

Segment 4
Crash-Test Dummies
1. At the time of the video, air bags had to protect a correctly seated, unbelted adult male dummy in a 30 mi/h crash. One change being proposed is to perform the crash tests using a variety of dummy sizes, including dummies representative of women and children. Another proposed change involves performing crash tests at 25 mi/h in addition to performing them at 30 mi/h.
2. The holes are needed so that the air bag can deflate quickly after it has deployed. These holes do not reduce the effectiveness of the air bag because they do not diminish the initial force exerted by the air bag.
3. During a crash, the occupants' bodies continue to move at the same velocity as the car was moving before it crashed. The unbalanced force that keeps each occupant from slamming into the windshield and dashboard is the force of the seat belt holding the person in the seat.

Segment 5
Smart Skin Sensors
1. People are most likely to be injured by air bags when they are sitting too close to where the air bag deploys.
2. The smart skin sensor can tell where the person's center of mass is, which indicates how the person is positioned in the seat.
3. The three options for the air bag are full deployment, soft deployment, or no deployment.
4. The sensor could send information regarding a passenger's position, causing the air bag to either deploy fully, softly, or not at all. The result would be the best protection possible for the passenger without causing injury.

Science in the News: Critical Thinking Worksheets Answer Key

Segment 6
Egg Drop Contest
1. One of the strategies used by students was to cushion the egg so that the force on the egg would be less when it hit the ground. The other strategy students used was to reduce the speed of the falling egg by maximizing air friction.
2. 1.2 kg·m/s downward
   ρ = (68 g)(1 kg/1000 g)(17 m/s)
   = 1.2 kg·m/s downward

Segment 7
Zero-Gravity Plane
1. The students are experiencing free-fall acceleration, meaning that they are accelerating from the force due to gravity.
2. The plane must be following a path similar to a parabolic curve, which slopes steeply upward to a maximum and then slopes back down.
3. When the plane levels off at the bottom of the dive, the students contact the floor of the plane because of their tendency to continue accelerating downward from the force due to gravity.

Segment 8
Circus Acrobats
1. The acrobats change the diameter of their bodies by bending slightly to decrease the length of their bodies.
2. It is easier to do a somersault in a tucked position. Doing a somersault in a straight position requires more energy because more inertia must be overcome.

Segment 9
Turbulent Flow
1. The jet rolls slightly and its path becomes unstable. The jet has this motion because it is flying through the air turbulence created by another jet.
2. Two advantages of using soap films to study turbulence are that the films provide a great visual and that they allow scientists to proceed quickly with experiments.

Segment 10
Wet Design
1. The water gun creates sudden, high spurts of water by releasing bubbles of high-pressure, compressed air.
2. The fountains that the company designs must be unique, artful, and entertaining.
3. Laminar flow is the regular, continuous, smooth movement of a fluid. Turbulent flow is fluid flow characterized by random fluctuations in velocity from point to point.

Segment 11
Energy-Saving House
1. All the appliances in the house can be powered by 1.7 kW of electrical energy because the house gets most of its energy from photovoltaic panels, which are powered by the sun. The energy generated by the photovoltaic panels can either be used, stored, or even sold to the power company.
2. In summer, heat is carried out through the underground water pipes. In winter, the heat pump draws energy from the water warmed in the ground.
3. The house is heated in winter because the sun shines through the south-facing windows in the afternoon. The overhang built above the south-facing windows helps keep the house from overheating in summer. This arrangement works both in summer and in winter because the angle of the sun changes as the seasons change. In summer, the sun is higher in the sky, so the overhang provides needed shade from the sun. In winter, the sun is lower in the sky, so the overhang does not prevent the sun from shining through the windows.
4. Power companies offer incentives to construct energy-saving homes because the companies would rather help people decrease their energy use than spend money meeting increasing energy demands.

Science in the News: Critical Thinking Worksheets Answer Key

Segment 12
Urban Heat Islands
1. Increased population results in more development and a larger urban area. This increased development and "urban sprawl" is what really causes cities to become heat islands. The cause could have been stated more precisely by saying, "More people require more buildings, roads, parking lots, etc., at the expense of green space, which contributes to more heat being reradiated by the area."
2. People are using lighter-colored materials when building new structures and either keeping existing green space or creating new green space.

Segment 13
Water-Cooled City
1. Lake Ontario remains cold in summer while the surrounding land becomes hot because water has a high specific heat, meaning that it takes a lot of energy transferred as heat to raise the temperature of 1 kg of water by 1 K.
2. The city of Toronto hopes to reduce the amount of electricity used for air conditioning by 90 percent. If they are successful, there will be much less air pollution caused by burning coal.

Segment 14
Virtual Practice Room
1. When the system was operating, the sound of the girl's playing became louder because the sound echoed (the sound waves reflected off the walls of the room instead of being absorbed by them as they were before).
2. Reverberations are repeated echoes. There would be more reverberations in a smaller arena than there would be in a large arena.
3. A school might install such practice rooms so that students studying music can experience the effect of playing in a large arena without the school actually having to build such an arena.

Segment 15
Japanese Telescope
1. A concave mirror must be used as the telescope's main mirror so that all the light is reflected to a main focal point in front of the mirror. Neither a convex mirror nor a flat mirror focuses light in front of the mirror.
2. Looking at distant stars and galaxies could yield clues about the origin of life because it is believed that the universe is expanding. When scientists observe distant stars and galaxies, they are actually seeing how these stars and galaxies looked many millions of years ago. (The farther away an object is, the older the light is that we see from that object.)

Segment 16
Color-Deficiency Lenses
1. The two kinds of light-sensitive cells in the retina of the eye are rods and cones, named for their shapes. Rods are extremely sensitive to dim light but do not distinguish colors. Cones are sensitive to bright light and color.
2. A more accurate description is that cones are able to absorb light at different wavelengths, corresponding to the colors blue, green, and red. This information is then sent to the brain for interpretation.
3. It is possible to see many different colors in many different shades because color vision results from the relative intensity of response of the three different kinds of cones. For instance, if all three kinds of cones respond with equal intensity, the color white is perceived.

Science in the News: Critical Thinking Worksheets Answer Key

Segment 17
Holograms
1. The lasers used in the laboratory are red, blue, and green. Three lasers are necessary so that different color combinations can be made to get true-color holograms.
2. The lasers are mounted on a floating table so that they are not affected by Earth's vibrations.
3. The Cartier company is interested in these full-color holograms because the company would be able to display exquisite pieces of jewelry without having to worry about the pieces getting stolen.

Segment 18
Fusion Lighting
1. The bulb shown in the video is smaller, does not have a filament, and has a unique shape that is similar to that of a lollipop.
2. The basis for the claim that sulfur lamps will last much longer than ordinary lamps is that sulfur lamps will not wear out. For example, there are no electrodes to evaporate and there is no mercury to interact with. Unlike ordinary lamps, sulfur lamps are completely inert.
3. If fusion lighting were used in American factories and in other American businesses, the cost of lighting (and therefore the cost of the products produced) would be reduced. As a result, America's competitive position internationally would improve because American businesses would be able to offer consumers cheaper products.
4. A plasma is a high-temperature gas consisting of ions, electrons, and neutral particles. The behavior of a plasma is dominated by the electromagnetic interactions between charged particles.

Segment 19
Edison's Lab
1. Edison's inventions brought light, sound, music, and motion pictures to people's lives in a way they had never experienced before.
2. Three of Edison's inventions were the alkaline battery, the phonograph, and the motion-picture camera.
3. Answers will vary but might include that this statement is false because technological advances are inevitable. For example, if Edison had not made these advances, someone else would have.
4. Edison's lab was the first industrial research lab in the United States. It helped set the standard for modern-day invention and research.

Segment 20
Eagle Electrocution
1. The wingspan of a large bird is large enough to complete an electric circuit, whereas the wingspan of a smaller bird is not.
2. It is believed that many of these deaths and injuries could be prevented by installing insulating sleeves over the wires and caps over the transformers where wires cross over utility poles.
3. Step-up transformers are used at or near the power plant to increase the voltage across power lines so that less energy is "lost" as heat due to the resistance of the transmission wires. Step-down transformers are used near homes and businesses to reduce the voltage across power lines so that the electric energy supplied is safer to use.

Science in the News: Critical Thinking Worksheets Answer Key

Segment 21
Magnetic Attractions
1. A pollutant may be bonded to a compound and then sent through magnets that selectively remove the pollutant while allowing the compound to pass through.
2. Paramagnetism is a kind of magnetism characteristic of materials weakly attracted by a strong magnet. Diamagnetism is a kind of magnetism characteristic of materials that line up at right angles to the magnetic field in which they are placed, therefore weakening the magnetic field. Bar magnets exhibit paramagnetic behavior. Graphite, plastics, grapes, habañero peppers, and crickets are just a few examples of objects that exhibit diamagnetic behavior.
3. Gravity pulls the object downward while repulsive magnetic forces push the object upward. Scientists are able to levitate objects by balancing these two opposing forces.

Segment 22
Atom Laser
1. The atom beam is a concentrated beam of matter, whereas a laser beam is a concentrated beam of light.
2. When the atoms are cooled to near absolute zero, their motion slows so that their wavelike behavior can be seen.
3. The atom laser may be used in space and sea navigation, in traffic lights, and to make the atomic clock even more precise. In the more distant future, the atom laser may even allow scientists and engineers to build better computer chips atom by atom.

Segment 23
Portable Solar Power
1. The solar photovoltaic generating system is particularly suited to this research facility because the facility is inhabited by only a small number of people who do not need much power. The system would probably not be as useful in northern California or in Oregon because there are many more people in these locations, and the system would not be able to keep up with the high demand for power.
2. It is believed that installing photovoltaic panels is a better idea than connecting the stations to power lines more than five miles away because it would cost $750,000 to connect the stations to the power lines. In addition, using the panels keeps the environmentally sensitive desert from being disturbed.
3. 45.5 A
   $I = (10.0 \text{ kW})(1000 \text{ W}/1 \text{ kW})/(220. \text{ V})$
   $= 45.5 \text{ A}$

Segment 24
Wisp of Creation
1. The video suggests that the elements hydrogen and helium were present after the big bang.
2. The farther away helium is found in the universe, the closer the helium is to the origin of the big bang and the more knowledge scientists can acquire about the early universe.
3. Dr. Davidsen characterizes "dark matter" as the matter that is not accounted for in stars. It is the matter that is not visible.